YOUR KNOWLEDGE HAS VALUE

- We will publish your bachelor's and master's thesis, essays and papers

- Your own eBook and book - sold worldwide in all relevant shops

- Earn money with each sale

Upload your text at www.GRIN.com and publish for free

Visha Rathod

Comparative Analysis for DNA Isolation from Jatropha curcas L.

GRIN Verlag

Bibliografische Information der Deutschen Nationalbibliothek:

Die Deutsche Bibliothek verzeichnet diese Publikation in der Deutschen National-
bibliografie; detaillierte bibliografische Daten sind im Internet über http://dnb.d-
nb.de/ abrufbar.

Imprint:

Copyright © 2010 GRIN Verlag GmbH
Druck und Bindung: Books on Demand GmbH, Norderstedt Germany
ISBN: 978-3-656-47034-2

This book at GRIN:

http://www.grin.com/en/e-book/230653/comparative-analysis-for-dna-isolation-
from-jatropha-curcas-l

COMPARATIVE ANALYSIS FOR DNA ISOLATION FROM *Jatropha curcas L.* (PROMISING BIODIESEL PRODUCING PLANT)

Project Report

By

Ms. Visha M. Rathod

Work carried out at

SHREE M. & N. VIRANI SCIENCE COLLEGE, RAJKOT
(Accredited at the "A" Level by NAAC & "STAR" College by MST-DBT)
DEPARTMENT OF BIOTECHNOLOGY
RAJKOT -360005
GUJARAT
INDIA

DECLARATION

I, hereby declare that the project entitled, *"Comparative analysis for DNA isolation from Jatropha curcas"* is my own work conducted under the supervision of Ms. Shivani Patel, Department of Biotechnology, Shree M. & N. Virani Science College. Rajkot.

I further declare that to the best of my knowledge the project doesn't contain any part of any work that has been submitted for the award of any degree either in this college or in any other college/institute/university without proper citation.

---------sd---------

Place: Rajkot

(Visha M. Rathod)

ACKNOWLEDGEMENT

Success is not the job of one person, it requires support of all. Firstly, I owe debt to God & my parents by whose blessings I am able to accomplish my work satisfactorily. I am grateful to Shree M. & N. Virani Science College. Rajkot for giving me an opportunity to carry out my work in this college.

A gesture of thankfulness & gratitude to my esteemed guide Ms. Shivani Patel, Department of Biotechnology, Shree M. & N. Virani Science College. Rajkot for their dynamic guidance, continuous encouragement, constant support and kind advice.

I also would like to convey my gratitude to **Ms. Shivani Patel**, Head, Department of Biotechnology, and Shree M. & N. Virani Science College. Rajkot, for her valuable suggestions during the course of study. Also, my hearty thanks to Mr. Nilkanth Faldu, Mr. K.P. Shenthil, Mr. Kantrao Saware, Mr. Raviranjan, Mr. Hiren Serathiya, Mrs. Priyanka Gandhi, Mrs. Shweta Bhatt, Ms. Hiral Padia, Ms.Vaishali Jasani and all the faculty members of VSC. These people have contributed a great deal to my work.

My sincere thanks and deepest appreciation to my laboratory staff Mrs. Mital Patel and Ms. Meghal Upadhyay for their technical help with instrumentation and for providing all the necessary requirement.

I would like to express my sincere thanks to Ms. Leena Ambasana for their invaluable help & advice throughout my work.

Words fail to express my gratitude and deepest appreciation to my ever loving parents, family members & all who have helped me by providing their indispensable aid, everlasting love, prayers, and the base they have paved for me without which it would have been impossible to complete my work.

(Visha M. Rathod)

INDEX

LIST OF TABLES

LIST OF FIGURES

ABBREVIATION

CTAB – Cetyl Trimethyl Ammonium Bromide

EDTA – Ethidium Dichoro Tetra Acetate

NaCl – Sodium Chloride

SDS – Sodium Dodecyl Sulphate

PCR – Polymerase Chain Reaction

PVP – Polyvinyl Pyrrolidone

RAPD – Random Amplified Polymorphic DNA

Taq – *Thermus aquaticus*

TE – Tris: EDTA

EtBr – Ethidium Bromide

Tris-HCL – Tris Hydrochloric acid

1. INTRODUCTION

Biotechnological approaches involve the exploitation of natural substances in production process. The utilization of various parts of *Jatropha curcas* potentially improving the economic situation of various tropical countries.

J. curcas is a drought-resistant shrub or tree belonging to the genus *Euphorbiaceae,* which is cultivated in Central and South America, South-east Asia, India and Africa. *J. curcas,* which can be easily propagated by cuttings, is widely planted as a hedge to protect field, as it is not browsed by cattle. Like many other *Jatropha* species *J. curcas* is a succulent that sheds its leaves during the dry season. Large green to pale-green leaves. Fruits are produced in winter, or there may be several crops during the year if soil moisture is good and temperatures are sufficiently high. It is well adapted to arid and semi-arid condition and often used for erosion control.

Cultivation is uncomplicated. *Jatropha curcas* grows in tropical and subtropical regions. The plant can grow in wastelands and grows on almost any terrain, even on gravelly, sandy and saline soils. *Jatropha curcas* has limited natural vegetative propagation and is usually propagated by seed. Propagation through seed (sexual propagation) leads to a lot of genetic variability in terms of growth, biomass, seed yield and oil content. Low seed viability and the recalcitrant nature of oil seeds also limit seed propagation. However, clonal techniques can help in overcoming these problems that hinder mass propagation of this tree-borne oilseed species. Vegetative propagation has been achieved by stem cuttings, grafting, and budding as well as by air layering techniques

Nuts of *J.curcas* Used as a contraceptive in South Sudan. Roots ashes are used as a salt substitute. The oil has been used for illumination, soap, candles, and the adulteration of olive oil. Bark Used as a fish poison. The latex of *J. curcas* has been found to be strongly inhibitory to watermelon mosaic virus. The latex has been also used to promote healing to wounds.

The first commercial application of *J. curcas* was reported from Lisbon, where the oil imported from Cape Verde was used for soap production and for lamps. The press cake was used as a fertilizer for potatoes. Even today, *J. curcas* is mainly cultivated for the production of oil as a fuel substitute. However, trans-esterification of the oil for use in standard diesel engines is gaining more importance then direct utilization of the oil in adapted engines. New techniques such as the enzyme-supported oil extraction and more efficient trans-esterification processes have been evaluated.

All parts of *J. curcas* have been used in traditional medicine and for Veterinary purposes for a long time. The oil has been used as a purgative, to treat skin diseases and to soothe pain such as that caused by rheumatism. Detection of the leaves has been used against coughs or as antiseptics after birth, and branches as chewing sticks. Recently the substances responsible for wound healing and anti-inflammatory effects have been isolated and characterized.

Various extracts from *J. curcas* seeds and leaves showed molluscicidal, insecticidal and fungicidal properties. Biotechnological process related to the exploitation of *J. curcas* include genetic improvement of the plant, biological pest control, enzyme supported oil extraction, anaerobic fermentation of the press cake and the isolation of anti-inflammatory substance and wound healing enzymes.

In India and Africa still various parts of *J. curcas* have been used in tradition medicine. In Mali the leaves are known as a treatment for malaria. The leaf decoction is applied externally for inflammation. The root decoction is drunk against pneumonia and syphilis. In Mexico the latex is used for fungal infection in the mouth and digestive problems of children.

J. curcas are, in general, toxic to humans and animals. Numerous feeding experiments with different animal species have demonstrated the toxicity of the seeds as well as of the oil and the press cake. The latex of *J. curcas* has been found to be strongly inhibitory to watermelon mosaic virus. The latex has been also used to promote healing to wounds. The use of *J. curcas* oil for the control of cotton insect pests seemed to be a promising alternative to hazardous chemicals. The lowest gas exhaust was obtained with *J. curcas*.

Limitation of *Jatropha curcas*

Seeds of *Jatropha curcas* are toxic and its cake cannot be used as a fodder. The low yields may have been caused by the fact that un-adapted provenances have been used. Being highly cross-pollinated crop, Ratanjyot within its species is genetically different. Large variation for oil percentage in the seed, from 25-40% has been observed in it.

2. REVIEW OF LITERATURE

1. *Jatropha curcas*:-

J. curcas is a drought-resistant shrub or tree belonging to the genus Euphorbiaceae, which is cultivated in Central and South America, south-east Asia, India and Africa. Large green to pale-green leaves. Fruits are produced in winter, or there may be several crops during the year if soil moisture is good and temperatures are sufficiently high.

Cultivation is uncomplicated. *Jatropha curcas* grows in tropical and subtropical regions. The plant can grow in wastelands and grows on almost any terrain, even on gravelly, sandy and saline soils. Currently the oil from *Jatropha curcas* seeds is used for making biodiesel fuel in Philippines and in Brazil.

Nuts of *J. curcas* Used as a contraceptive in South Sudan. Roots ashes are used as a salt substitute. The oil has been used for illumination, soap, candles, and the adulteration of olive oil. Bark Used as a fish poison. The latex of *J. curcas* has been found to be strongly inhibitory to watermelon mosaic virus. The latex has been also used to promote healing to wounds.

2. DNA isolation from *J. curcas* plant:-

Non-renewable hydrocarbons are being used as the major energy source, which are mainly responsible for global warming. Biofuels and bioenergy encompass a wide range of alternative source of energy of biological origin. Plant based fuels create a better balance between the formation and consumption of CO_2; decrease particulate matter, co, unburnt hydrocarbon and so_2 emission into the atmosphere. In this regard, biodiesel derived from the seed-oil of *J. curcas* is fast emerging as an available alternative to fossil fuels and has a desirable physiochemical activity, even superior to diesel. The by-products, obtain while preparing biodiesel, have industrial application.

The deoiled cake is used as fertilizer and in biogas production. Almost all parts of the plant are used in medicine. Hence, there is a need to identify high yielding clones of *J. curcas* for its further improvement and large scale cultivation.

DNA isolation is a routine procedure to collect DNA for subsequent molecular or forensic analysis. There are three basic and one optional step in a DNA extraction:

1. Breaking the cells open commonly referred to as cell disruption or cell lysis, to expose the DNA within. This is commonly achieved by grinding or sonicating the sample.
2. Removing membrane lipids by adding a detergent.
3. Removing proteins by adding a protease (optional but almost always done).
4. Precipitating the DNA with an alcohol — usually ice-cold ethanol or isopropanol. Since DNA is insoluble in these alcohols, it will aggregate together, giving a pellet upon centrifugation. This step also removes alcohol-soluble salt.

3. Detecting DNA

Measuring the intensity of absorbance of the DNA solution at wavelengths 260 nm and 280 nm is used as a measure of DNA purity. DNA absorbs UV light at 260 and 280 nanometers, and aromatic proteins absorb UV light at 280 nm; a pure sample of DNA has the 260/280 ratio at 1.8 and is relatively free from protein contamination. A DNA preparation that is contaminated with protein will have a 260/280 ratio lower than 1.8.

A_{260} = measure concentration of the DNA (1.0 A_{260} = 50µg/ml)

A_{260}/A_{280} = Estimate DNA purity

4. Restriction digestion of DNA:-

Recombinant DNA technology is a modern methodology in the field of molecular biology to manipulate genes. It includes enzymes to cut DNA. The restriction endonuclease cut at the interior part of the DNA. These enzymes found in bacteria and in vivo are involved in recognition and destruction of foreign DNA. Invading phage

DNA for instance will be restricted by such enzymes. The bacteria protect their own DNA by modification process.

The essential feature of restriction endonuclease action is that the enzyme recognizes a particular sequence of bases. Bacteria are constantly attacked by bacteriophages to protect themselves; bacteria have developed defense mechanism in the form of enzyme known as restriction endonuclease.

5. RAPD analysis:-

RAPD stands for Random Amplification of Polymorphic DNA. It is a type of PCR reaction, but the segments of DNA that are amplified are random. The scientist performing RAPD creates several arbitrary, short primers (8-12 nucleotides), then proceeds with the PCR using a large template of genomic DNA, hoping that fragments will amplify.

RAPD analysis requires only a small amount of genomic DNA and can produce high levels of polymorphism and may facilitate more effective diversity analysis in plants (Williams et al., 1990). RAPD analysis provides information that can help to define the distinctiveness of species and phylogenetic relationships at molecular level (Subramanyam et al. 1901).

RAPD is an inexpensive and rapid method not requiring any information regarding the genome of the plant and has been widely used to ascertain genetic diversity in several plants

How It Works:-

Unlike traditional PCR analysis, RAPD does not require any specific knowledge of the DNA sequence of the target organism: the identical 10-mer primers will or will not amplify a segment of DNA, depending on positions that are complementary to the primers' sequence. For example, no fragment is produced if primers annealed too far apart or 3' ends of the primers are not facing each other. Therefore, if a mutation has occurred in the template DNA at the site that was previously complementary to the primer, a PCR product will not be produced, resulting in a different pattern of amplified DNA segments on the gel.

Limitations of RAPD:-

> Nearly all RAPD markers are dominant, i.e. it is not possible to distinguish whether a DNA segment is amplified from a locus that is heterozygous (1 copy) or homozygous (2 copies). Co dominant RAPD markers, observed as different-sized DNA segments amplified from the same locus, are detected only rarely.
> PCR is an enzymatic reaction, therefore the quality and concentration of template DNA, concentrations of PCR components, and the PCR cycling conditions may greatly influence the outcome. Thus, the RAPD technique is notoriously laboratory dependent and needs carefully developed laboratory protocols to be reproducible.
> Mismatches between the primer and the template may result in the total absence of PCR products well as in a merely decreased amount of the product. Thus, the RAPD results can be difficult to interpret.

3. MATERIALS AND METHODS

Plant material

Nine DNA extraction methods are examined on young and mature leaf samples of *Jatropha curcas*. Fresh leaves of *J. curcas* were collected from our experimental field located at Shree M. & N. Virani Science College, Rajkot, Gujarat, India, and brought to the laboratory. Leaves were harvested fresh before DNA isolation. The DNA was extracted from fresh leaves on the same day using different method to obtain high quality intact DNA.

For DNA isolation different nine protocols were examined, but from those only six protocols had given positive result. These nine methods and their chemical requirements are given below.

DNA extraction protocol A (modified - Doyle and Doyle, 1990)

Chemicals

- Extraction buffer: 2% CTAB, 100 mM Tris/HCl, pH 7.5, 1.4 M NaCl, 2% polyvinylpyrrolidone (PVP)-40, 20 mM ethylene diamine tetra acetic acid (EDTA), pH 8.0. Add 20 µL/mL β-mercaptoethanol immediately prior to use.
- Chloroform: isoamylalcohol 24:1 (CIA).
- Isopropanol, 70% ethanol.
- TE-RNase solution: 10 mM Tris-HCl, 1 mM EDTA, pH 8.0, 10 mg/mL RNase.

Protocol

1. 500 mg of leaf sample was crushed in mortal and pestle with the help of extraction buffer A and the crushed samples were equally distributed in the eppendroff tubes.
2. To ground sample, add 1 mL extraction buffer and incubate samples for 1 h at 60°C in boiling water bath with occasional swirling.
3. Cool samples at room temperature, and centrifuge at 14000 rpm for 10 minutes.
4. Supernatant was collected in fresh vial and add 600 µL CIA and mix gently for 5 min.
5. Then the sample was centrifuged for 15 min at 12000 rpm at 4°C. Transfer the supernatant to a new tube and add equal volume of Isopropanol. Mix gently and incubate at -20°C overnight.
6. Centrifuge for 15 min at 14000 rpm. The tubes were allowed to drain and Wash the pellet with 70% ethanol and allowed to dry.
7. Dry the DNA and dissolve the pellet in 20-50 µL (according to your quantity) TE-RNase solution. Incubate for 1 h at 37°C in dry bath
8. Equal amount of Isopropanol were added and samples were incubated at -20°C for 2 hours.

9. Samples were centrifuged at 14000 rpm for 15 minutes at 4°C in cooling centrifuge.
10. The tubes were allowed to drain and dried at room temperature for 20-30 minutes, and then DNA was resuspended in 50µl of Tris-EDTA buffer.

DNA extraction protocol B (Jobes et al., 1995)

Chemicals

- Extraction buffer: 100 mM sodium acetate, pH 4.8, 100 mM EDTA, pH 8.0, 500 mM NaCl, 10 mM dithiothreitol (DTT), 2% PVP, pH 5.5. Add 100 µg/mL proteinase K immediately prior to use.
- 20% sodium dodecyl sulfate (SDS) solution.
- 5 M potassium acetate, 8 M LiCl, 5 M NaCl.
- Isopropanol, absolute and 70% ethanol.
- TE buffer: 10 mM Tris-HCl, 1 mM EDTA, pH 8.0.

Protocol

1. 500 mg of leaf sample was crushed in mortal and pestle with the help of extraction buffer B and the crushed samples were equally distributed in the eppendroff tubes.
2. To each ground leaf sample, add 1 mL extraction buffer and incubate sample for 1 h at 60°C with occasional swirling.
3. Samples were cooled at room temperature, and centrifuged at 12000 rpm for 10 minutes at 4°C
4. Transfer the supernatant to a new tube and add 1/3 volume of 5 M potassium acetate. Mix gently and incubate for 30 min at -20°C.
5. Centrifuge for 10 min at 14000 rpm. Transfer the supernatant to new tube and add 0.6 volume of Isopropanol. Mix gently and incubate at -20°C overnight.
6. Centrifuge for 15 min at 14000 rpm. The tubes were allowed to drain and Wash the pellet with 70% ethanol and allowed to dry.
7. Dissolve the pellet in autoclaved Deionized water. Add 0.5 volume of 5 M NaCl and mix well. Add two volumes of cold absolute ethanol and incubate for 30 min at -20°C.
8. Centrifuge for 10 min at 14000 rpm. Dissolve the pellet in autoclaved deionized water. Precipitate RNA with 1/3 volume of cold 8 M LiCl (final concentration 2 M) and incubate at -20°C for at least 1 h.
9. Recover RNA by centrifugation for 15 min at 14000 rpm and carefully transfer supernatant to a new tube. Precipitate DNA by adding 0.6 volume of Isopropanol and incubate for 1 h at -20°C.
10. Centrifuge for 10 min at 14000 rpm. Wash the pellet with 70% ethanol and was allowed to dry.
11. Dried DNA was dissolved in 30µl of TE buffer.

DNA extraction protocol C (Dellaporta et al., 1983)

Chemicals

- Extraction buffer: 10% SDS, 50 mM Tris/HCl, 100 mM NaCl, 10 mM EDTA, pH 8.0. Add 20 µL/mL β-mercaptoethanol and approximately 0.01 g PVP-40 immediately prior to use.
- 5 M potassium acetate, 3 M sodium acetate.
- Isopropanol, absolute and 70% ethanol.
- TE-RNase solution, TE buffer.

Protocol

1. 500 mg of leaf sample was crushed in mortal and pestle with the help of extraction buffer C and the crushed samples were equally distributed in the eppendroff tubes.
2. To each ground leaf sample, add 1 mL extraction buffer and incubate sample for 1 hour at 60°C with occasional swirling.
3. Add 300 µL 5 M potassium acetate and mix gently. Incubate sample for 20 min on ice.
4. Centrifuge for 10 min at 14000 rpm. Transfer the supernatant to a new tube and add equal volume of Isopropanol. Mix gently and incubate for 1 h at -20°C.
5. Centrifuged at 14000 rpm for 15 min. Wash the pellet with 70% ethanol.
6. Dry the DNA and dissolve the pellet in 20-50 µL TE-RNase solution. Incubate for 1 h at 37°C in dry bath.
7. Precipitate the DNA by adding 10% volume of 3 M sodium acetate and two volumes of absolute ethanol. Incubate at -20°C overnight.
8. Samples were Centrifuged at 14000 rpm for 15 min at 4°C and tubes were allowed to drain and dry at room temperature for 20-30 minutes.
9. Dry the DNA and dissolve the pellet in 20-50 µL TE buffer.

DNA extraction protocol C2

Chemicals

- Extraction buffer: 1M Tris/HCl, pH 8, 10% SDS, 5 M NaCl, 4% polyvinylpyrrolidone (PVP)-40, 0.5 M ethylene diamine tetra acetic acid (EDTA), pH 8.0. Add 20 µL/mL β-mercaptoethanol immediately prior to use.
- Chloroform: isoamylalcohol 24:1 (CIA).
- Isopropanol, 70% ethanol.
- 1.5 M Na acetate, 1 X TE buffer

Protocol

1. 500 mg of leaf sample was crushed in mortal and pestle with the help of extraction buffer C2 and the crushed samples were equally distributed in the eppendroff tubes.
2. To each ground leaf sample, add 1 mL extraction buffer and incubate sample for 20 min at 65°C with occasional swirling.

3. Centrifuge for 10 min at 10,000 rpm. Transfer the supernatant to a new tube and add 500 µl sodium acetate solution mix it well.
4. Centrifuge for 10 min at 10,000 rpm. Transfer the supernatant to a new tube and add equal volume of CI(chloroform isoamylalcohol) solution. Mix it well and Centrifuge for 10 min at 10,000 rpm at 4°C.
5. Repeat above step until supernatant become transparent.
6. After centrifugation pipette out upper aqueous phase into a new tube and add equal amount of isopropanol to precipitate the DNA.
7. Incubate the tube at -20°C for 15 min.
8. After the incubation centrifuge at 14,000 rpm for 15 min at 4°C temp.
9. After the centrifugation discard the supernatant and wash the pellet with 70% ethanol.
10. Discard the ethanol and air dry the pellet.
11. Dissolve the pellet into 30 µl 1 X TE buffer.
12. Run the DNA onto a 1% agarose gel and observe under the U.V. light for the presence of plant genomic DNA.

DNA extraction protocol G

Chemicals

- Extraction buffer: 2% SDS, 100 mM Tris/HCl, 500 mM NaCl, 20 mM EDTA, pH 8.0. Add 0.1% β-mercaptoethanol and 7M urea
- Phenol: chloroform: isoamylalcohol (25:24:1)
- 3M sodium acetate
- Isopropanol, absolute and 70% ethanol.
- TE-RNase solution, TE buffer.

Protocol

1. 500 mg of leaf sample was crushed in mortal and pestle with the help of extraction buffer G and the crushed samples were equally distributed in the eppendroff tubes.
2. To ground sample, add 1 mL extraction buffer and incubate samples for 1 h at 60°C in boiling water bath with occasional swirling.
3. Cool samples at room temperature, and centrifuge at 14000 rpm for 10 minutes.
4. Supernatant was collected in fresh vial and add equal amount of phenol: chloroform: isoamylalcohol (25:24:1) was added and mix by shaking.
5. Centrifuge for 20 min at 14000 rpm at room temperature.
6. Aqueous phase was collected in fresh vial and 600µl of chloroform: isoamylalcohol was added and was mix by gently for 5 min.
7. Then the sample was centrifuged for 15 min at 12000 rpm at 4°C. Transfer the supernatant to a new tube and add 0.1 volume of 3M sodium acetate and 0.7 volume of Isopropanol. Mix gently and incubate at room temperature for 30 minutes.
8. Centrifuge for 15 min at 14000 rpm. The tubes were allowed to drain and Wash the pellet with 70% ethanol and allowed to dry.
9. Dry the DNA and dissolve the pellet in 20-50 µL (according to your quantity) TE-RNase solution. Incubate for 1 h at 37°C in dry bath

10. Equal amount of Isopropanol were added and samples were incubated at -20°C for 2 hours.
11. Samples were centrifuged at 14000 rpm for 15 minutes at 4°C in cooling centrifuge.
12. The tubes were allowed to drain and dried at room temperature for 20-30 minutes, and then DNA was resuspended in 50µl of Tris-EDTA buffer.

DNA extraction protocol H

Chemicals

- Extraction buffer: contained 2% CTAB, 100 mM Tris-HCl, 3.5 M NaCl, 20 mM EDTA, 0.2M β-mercaptoethanol and 2% PVP
- Chloroform: isoamylalcohol 24:1 (CIA).
- Phenol: chloroform: isoamylalcohol (25:24:1)
- 5 M NaCl, 3 M sodium acetate
- Isopropanol, 70% and 80% ethanol.
- TE buffer(10 mM Tris-HCl, 1 mM EDTA, pH 8.0)
- 10mg/ml RNase solution

Protocol

1. 500 mg of leaf sample was crushed in mortal and pestle with the help of extraction buffer H and the crushed samples were equally distributed in the eppendroff tubes.
2. To ground sample, add 1 mL extraction buffer and incubate samples for 90 min at 65°C in boiling water bath with occasional swirling.
3. Cool samples at room temperature, and centrifuge at 14000 rpm for 10 minutes.
4. Supernatant was collected in fresh vial and add equal amount of CIA and mix gently for 5 to10 min to form an uniform emulsion.
5. The mixture was centrifuged at 8000 rpm for 8 min at RT.
6. Chloroform: isoamylalcohol extraction step was repeated again.
7. The aqueous phase was pipette out gently, avoiding the interface.
8. To the above solution, 5M NaCl and 0.6 volume of Isopropanol of the total solution was added and incubated at RT for 1 hour.
9. To the above solution, two volumes of 80% ethanol was added and incubated for 10 min at RT for DNA precipitation.
10. After incubation the mixture was centrifuged at 10,000 rpm for 15 min.
11. The pellet was washed with 70% ethanol, dried and resuspend in 200µl of TE buffer.

DNA Quantification

The yield of extracted DNA was quantified by spectrophotometer at 260 nm. After diluting the DNA two hundred times in DNase free water, it was quantified by taking the optical density (OD) at λ 260 with a spectrophotometer. The purity of Genomic DNA was determined by the A260/A280 absorbance ratio. The quality was also examined by running the extracted DNA samples on 0.8 % agarose gel stained with 10 mg/ml ethidium bromide in 1×TBE (Tris base, Boric acid, EDTA) buffer. The gel was visualized and photographed under UV light.

- A_{260} = measure concentration of the DNA (1.0 A_{260} = 50µg/ml)
- A_{260}/A_{280} = Estimate DNA purity

Restriction Digestion of Genomic DNA

The extracted DNA was digested by two enzymes Eco-R1 and Bam-H1 separately.

Chemicals:
Deionized water, lambda DNA, assay buffer, restriction enzyme, bromo phenol blue, 1% agarose gel, EtBr

Protocol:

1. Take sterile microfuge tube & add the reaction components in a order given in the table.
2. Table for reaction component:-

Table1: Reaction component of RE digestion

NO	COMPONENT	AMOUNT
1	Deionized water	39 µl
2	Assay buffer	5 µl
3	Lambda DNA	5 µl
4	R.E.	1 µl

3. After adding all the reaction components mix the content by gently tapping tube & incubate the tube at 37°C for 1 hour in a dry bath.
4. Add 3 µl of bromo phenol blue to stop the reaction.
5. Run the DNA on 1% agarose gel and observe the band pattern under the U.V. light.

RAPD Analysis:-

The PCR amplification was carried out in biotech department, M. & N. Virani Sci. College. RAPD analysis of DNA was done by primer no. 13th, whose sequence is given below.

Primer no.	Sequence of primer
Primer 13	5' ACCAGGTCAC 3'

RAPD PCR Reaction mixture

Table2: reaction component of RAPD analysis

COMPONENTS	AMOUNT
D/W	7.0 µl
10x Assay buffer	1.25 µl
Mgcl2	1.0 µl
Primer	1.0 µl
Dntp	1.0 µl
Taq polymerase	0.25 µl
DNA (100ng/100 µl)	2.0 µl

RAPD PCR Program:-

Amplification was done using thermal cycler with following program:

Table 3: PCR condition for RAPD analysis

STEPS	TEMPERATURE	TIME
Initial denaturation	95 °C	1 min
Denaturation	95 °C	30 sec
Annealing	32 °C	1 min
Extension	72 °C	1 min
Go to	2 repeat 39 cycle	
Final Extension	72 °C	5 min
Hold	4 °C	∞

4. RESULTS AND DISCUSSION

DNA isolation

Quality and quantity of extracted DNA is tested in spectrophotometer methods, electrophoresis on 0.8% agarose gel gave discrete single band of pure DNA. By spectrophotometer procedure, absorption of double-stranded DNA in wavelength of 260 nm. In due to the fact, when the ratio of absorption amount resulted in 260 nm to the result of 280 nm is between 1.0 and 1.4, it shows the most absorption is done by nucleic acids and therefore extracted DNA is well-qualified and its purity is acceptable. So these results confirm that extracted DNA executed by D V N Sudheer Method from young and mature leaves possess better quality and quantity in compare with the other methods. (fig.4) Using 0.5 g leaf material in this method gained the most DNA extraction 6350 ng mL-1 of young leaves. Though the quality of extracted DNA by Doyle and Doyle method was agreeable, its value was less than what has been extracted by the D V N Sudheer: only 3350 ng mL-1. (fig.1)DNA extractions by Jobes *et al.* (1998) and also Dellaporta *et al.* (1983) methods produced superfluous materials so that the ratio of absorption in 260 nm to 280 nm equaled a range between 1.1 and 1.3, which DNA electrophoresis confirmed as well. In comparison with the other methods, for both young and mature leaves, the amount of extracted DNA by the D V N Sudheer was higher than Doyle and Doyle's. Its quality was much better than the result of other methods too.

Electrophoresis on agarose gel for DNAExtracted from nine different protocols. But we get result in only six protocols *Left to right*:
(A). Doyle and Doyle (1990), (B). Jobes *et al.*(1995), (C). Dellaporta *et al.* (1983), (C2). Rapid extraction,
(G). Modified SDS method, (H). D V N Sudheer

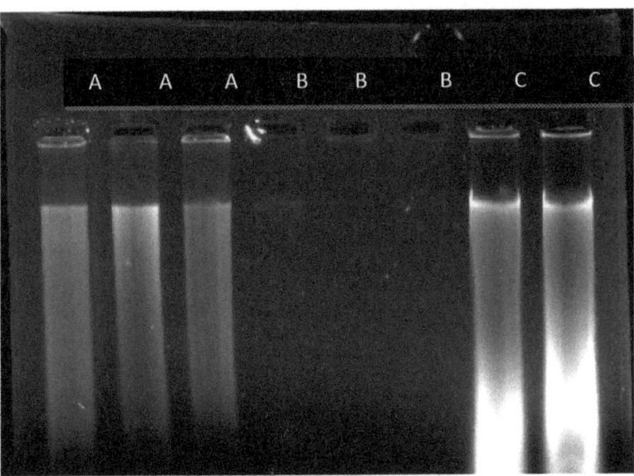

Figure 1. Purified genomic DNA of Jatropha curcas accessions isolated using A. Doyle and Doyle (1990). B. Jobes *et al.*(1995) C. Dellaporta *et al.* (1983)

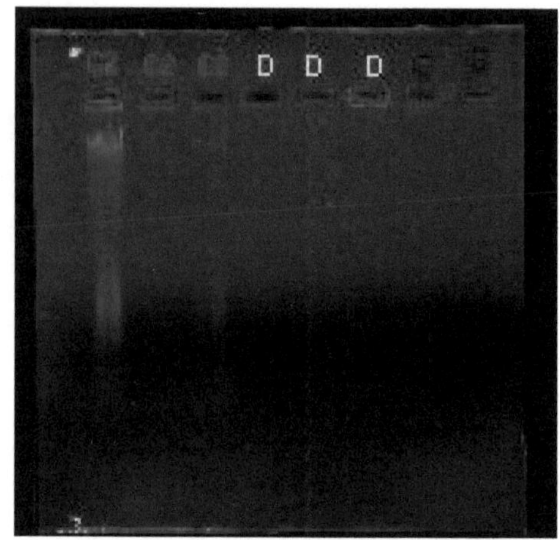

Figure2. Purified genomic DNA of Jatropha curcas accessions isolated using C2. Rapid extraction, In protocol D and E we didn't get result.

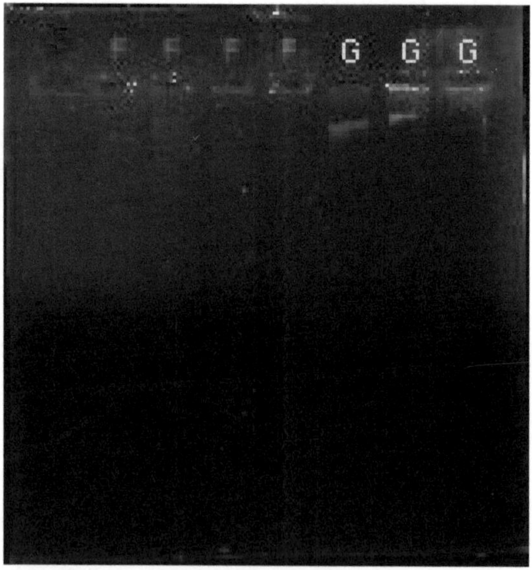

Figure3. Purified genomic DNA of Jatropha curcas accessions isolated using protocol E, protocol F, protocol G. In protocol E and F we didn't get result.

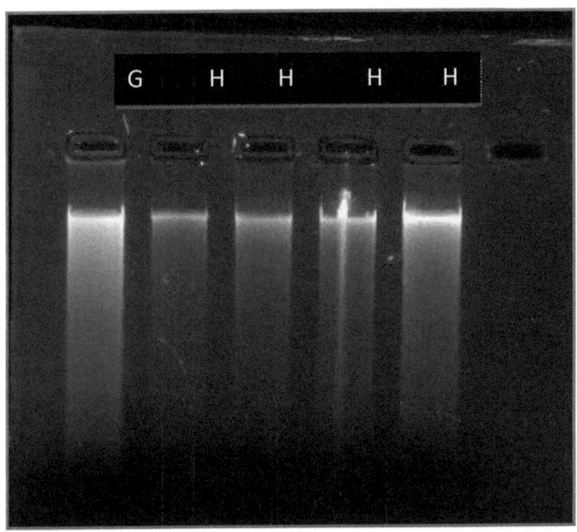

Figure 4. Purified genomic DNA of Jatropha curcas accessions isolated using protocol G and protocol H. we get very good intact DNA band in protocol H.

QUANTIFICATION OF DNA:-

By spectrophotometer procedure, absorption of double-stranded DNA in wavelength of 260 nm. In due to the fact, when the ratio of absorption amount resulted in 260 nm to the result of 280 nm is between 1.0 and 1.4, it shows the most absorption is done by nucleic acids and therefore extracted DNA is well-qualified and its purity is acceptable. So these results confirm that extracted DNA executed by D V N Sudheer Method from young and mature leaves possess better quality and quantity in compare with the other methods. Concentration of DNA and purity of DNA extracted by different methods are given in table 4.

Table 4:- It shows concentration of DNA and purity DNA isolated by different method.

Protocol	O.D. at 260 nm	O.D. at 280 nm	conc^n µg/ml	Purity	Purity (%)
A	0.067	0.059	3.35	1.135	63.05%
B	0.061	0.054	3.05	1.129	62.72%
C	0.124	0.092	6.20	1.347	74.83%
C2	0.059	0.054	2.95	1.092	60.55%
G	0.068	0.060	3.40	1.133	62.77%
H	0.127	0.085	6.35	1.494	83.00%

Restriction Digestion:-

By performing six methods for DNA isolation, we get very god quality of DNA which is not susceptible to restriction enzymes. Here DNA is not affected by restriction enzyme Eco-R1 and Bam-H1 so, we didn't get restriction bands.

It means Quality of extracted DNA is very good, which can be used in future for more research work.

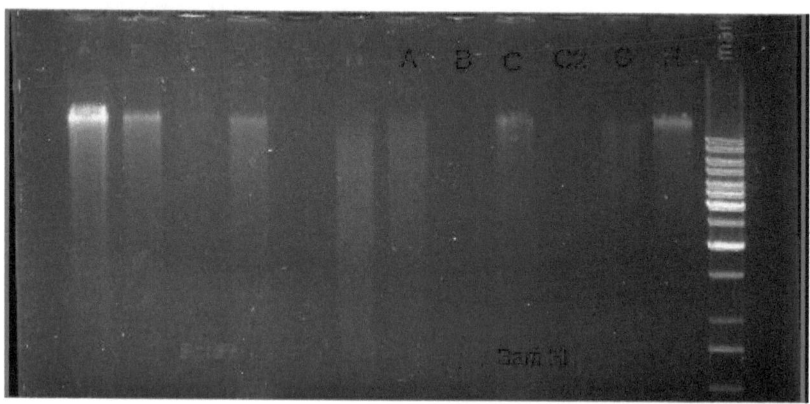

Fig.5 Restriction digestion of genomic DNA by 2 enzymes Eco R1 and Bam H1.

RAPD PCR Analysis:-

By performing six methods for DNA isolation, we got very good quality of DNA, which can be used for further amplification. It has seen very good polymorphism by RAPD analysis. It will prove very useful for further molecular experiments on *Jatropha curcas.*

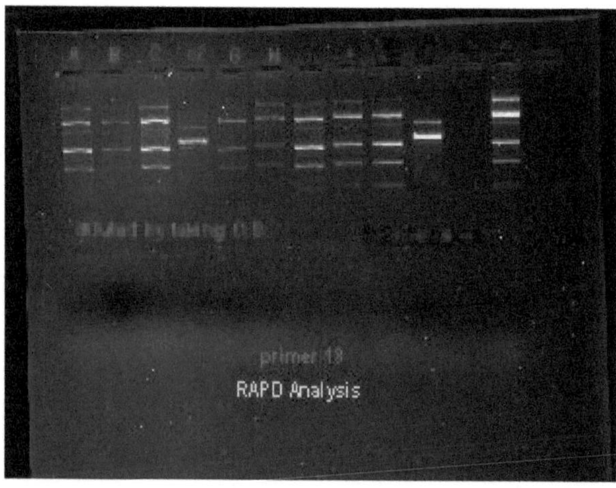

Fig. 6 Electrophoretic pattern of DNA of Jatropha curcas accessions amplified with RAPD   primer 13

DISCUSSION

DNA isolation

Different frequently used methods of DNA isolation from plants did not yield good quality DNA from *Jatropha curcas* as *J.curcas* leaves contain many alkaloids, protein, polysaccharide. These compounds pose difficulties in recovery and further processing of supernatants during normal isolation of DNA from *Jatropha*. The DNA obtained was contaminated with pigments and some of them were highly viscous. Due to contamination of the polysaccharide and protein with DNA the extracted sample, upon electrophoresis, gave fire-type bands, uneven migration, and often remained in the wells during electrophoresis. The unsuccessful attempts for extraction of good quality DNA from *J. curcas* with the available protocols made to search for the new extraction protocol. Good quality of DNA was isolated using standardized method of DNA isolation from *Jatropha* reported by D V N Sudheer.

RAPD based assessment of genetic diversity

Despite its potential as an alternative to fossil fuel, *Jatropha* is not being fully exploited due to lack of quality planting material for cultivation. Hence, there is a need to identify high yielding clones of *J. curcas* for its further improvement and large scale cultivation. Improvement of this plant is thus of prime importance, where in, maximum desirable traits can be brought together in one plant. Therefore, necessity to study genetic variability in *Jatropha curcas* has given immense important for selection of superior genotypes with high seed yield and more oil content in their seeds. DNA based assays such as Random Amplified Polymorphic DNA (RAPD) is most widely used tool for assessment of the genetic variation in the medicinal plants.

Advances in the field of molecular biology have provided many tools for studying the diversity in genome level to get phylogenetic relationship among different species. Out of many PCR based fingerprinting techniques RAPD emerged as very useful and efficient methods for analyzing the molecular diversity due to ease of working and efficiently utilized for studying the divergence between different plant species, within the species and varieties. RAPD analysis is proved to be rapid, inexpensive and useful tool for assessing phylogenetic relationship among species.

SUMMARY OF REVIEW

The most essential principle in the modern molecular biology is extraction of DNA with a desirable quantity and quality, and this achievement and complete DNA sequencing of genome could be the main necessity for every genetic study. DNA extraction is obviously difficult because of negative effects obtained from carbohydrates, tannins, poly phenols and proteins. Therefore, the ideal method is the one, which highly reduces the presence of these materials. In the present study, four special methods are chosen to be compared: (1) modified - Doyle and Doyle, 1990, (2) Jobes et al., 1995, (3) Dellaporta et al., 1983, and (4) Rapid extraction protocol, (5)modified SDS method, (6) D V N Sudheer Pamidimarri et al. Both young and mature leaves of *Jatropha curcas* are used in the procedure. The quality and quantity of extracted genomic DNA gained from these methods are deliberated by means of spectrophotometry, electrophoresis in 1% agarose gel and PCR. In this regard, application of D V N Sudheer Pamidimarri et al. is chosen for young and mature leaves. The most value of qualified DNA is extracted from young leaf tissue by 6350 ng/ml. According to the data, D V N Sudheer Pamidimarri et al. is recommended for DNA extraction from young leaves of *Jatropha curcas*.

5. CONCLUSION

Jatropha curcas plant is useful in various fields like medicine, biodiesel, latex. So, by comparative study for DNA isolation, It will prove very useful for further molecular experiments on Jatropha curcas. From the above result it can be concluded that good Quality and Quantity of DNA can be obtained by D. V. N. Sudheer method. The quality of DNA is very good because it is not susceptible to restriction enzyme and also it shows good polymorphism by RAPD analysis. Such DNA will be helpful for further research work in molecular biology.

6. REFERENCES

1. Anna Maria Pirttila Et Al., DNA Isolation Method For Medicinal & Aromatic Plants, Plant Molecular Biology Reporter 19.
2. Dharman Dhakshanamoorthy1 And Radhakrishnan Selvaraj, Extraction Of Genomic DNA From *Jatropha Sp.* Using Modified Ctab Method
3. D V N Sudheer, A Simplified Method For Extraction Of High Quality Genomic DNA From Jatropha Curcas For Genetic Diversity And Molecular Marker Studies, Indian Iournal Of Biotechnology, Vol 8, April 2009, Pp 187-192.
4. E.Nalini & N. Jawali Et Al, A Simple Method For Isolation Of DNA From Plants Suitable For Long Term Storage & DNA Marker Analysis, BARC Newsletter Issue No. 249.
5. G. M. Gubitz Et Al. , Exploitation Of The Tropical Oil Seed Plant *Jatropha Curcas* L. ,
Bioresource Technology 67 (1999) 73-82.

6. Mitra Alaey1, Rouhangiz Naderi Et Al, Comparing Study Between Four Different Methods Of Genomic DNA Extraction From *Cyclamen Persicum* Mill, International Journal Of Agriculture & Biology 1560–8530/2005/07–6–882–884

7. R.A. Ribeiro & M.B. Lovato, Comparative Analysis Of Different DNA Extraction Protocols In Fresh & Herbarium Specimens Of The Genus *Dalberia,* Page (173-187)

8. Sadasivam, S. & A. Manickam, Biochemical Methods For Agricultural Sciences, Wiley Estern Limited, New Delhi, Pp. 150 -151.

9. S. R. Thimmaiah, Std. Methods Of Biochemical Analysis, Page (155-157) (178-183)

ANNEXURE-I

A. Solution, Chemicals And Reagents Used For DNA Extraction
1. Tris: HCL buffer (pH-8.0)-1M solution.

12.11gm Tris salt was dissolved in sterile distilled water and pH adjusted to 8.0 using 1N HCL. The volume was made up to 100ml using sterile distilled water and the solution was autoclaved prior to usage.

2. Ethylene Diamine Tetra Acetic Acid (EDTA)-0.5M solution

18.62g EDTA was dissolved in sterile distilled water, the pH of which was raised to 8.0 by addition of 1 M NaOH. The solution was stirred vigorously with a megnatic stirrer and after complete dissolution of the solute, the volume was made up to 100ml with sterile distilled water, the solution was autoclaved prior to usage.

3. Sodium Chloride (NaCl)- 5M solution

29.2g was dissolved in sterile distilled water and volume was made up to 100ml with sterile distilled water, the solution was autoclaved prior to usage.

4. Cetyl Trimethyle Ammonium Bromide (CTAB) – 20% s0lution

20g CTAB was dissolved in sterile distilled water and volume was made up to 100ml with sterile distilled water. the solution was autoclaved prior to usage.

5. DNA extraction buffer

DNA extraction buffer was prepared a fresh as and when necessary using stock solutions of the individual components.

Component	Concentration of stock solution	Concentration of working solution	Volume of stock taken for 100ml solution (ml)
CTAB	20%	2%	10
NaCl	5M	3.5M	70
EDTA	0.5 M	20mM	4
Tris	1M	100mM	10
B Mercaptoethanol	2%	0.2%	0.3
PVP		2%	2gm
Sterile water			5.7

β Mercaptoethanol was added at last after adding extraction buffer to ground plant samples.

6. Isopropanol

7. Sodium acetate-3 M solution

40.824g sodium acetate was dissolved in sterile distilled water and volume was made up to 100ml with sterile distilled water. the solution was autoclaved prior to usage.

B. DNA Purification
1. RNase A (10mg/ml solution)

RNase	10mg
Tris HCL (pH-7.5)	10mM
NaCl	1.5mM

Sterile water was added to make volume to 1ml. The solution was heated at 100° C for 15 min and then stored at -20° C.

C. Solvent for DNA
1. Tris: EDTA (TE) buffer (10mM)

10ml of Tris (1M) buffer and 0.2 ml of EDTA (0.5M) was mixed with sterile water and volume made up to 100ml with sterile distilled water. The solution was autoclaved prior to usage

D. Gel Electrophoresis
1. 50 x TAE (Tris/Acetate/EDTA)

Tris base	242gm
Glacial Acetic Acid	57.1ml
0.5M EDTA (pH-8.0)	100ml
Distilled water	to 1 liter

The pH of diluted sample is approx 8.5

2. 10x TBE (Tris/Borate/EDTA)

Tris base	108gm
Boric acid	55gm
0.5M EDTA	40 ml
Distilled water	to 1 liter

3. Agarose gel (1.5%)

1.5gm of agarose was added in 100ml 1xTAE buffer, the content was mixed thorouly and boiled for 1-2min to dissolve the content. The mixture was cooled down to 40° C. The molten gel was casted in a tray with a comb containing teeth to produce wells.

4. Ethidium Bromide (10mg/ml)

10 mg of ethidium bromide was dissolved in sterile water and volume made up to 1ml. The solution was stored in an amber colored bottle at 4° C

5. Loading dye (10 x solution)

Bromophenol Blue	0.25%
Xylene cyanol FF	0.25%
Glycerol	50%
TAE	1x

Sterile water was added to the above compounds to make volume to 100ml.

ANNEXURE-II

Reagent/Chemicals – Source

- All the chemicals used in this study were of molecular biology grade procured either from HIMEDIA / QUALIGEN.
- The components of PCR reaction viz, dNTP's, PCR buffer, Taq DNA polymerase, $MgCl_2$ etc. were purchased from Genie, Bangalore.
- All the instruments used were of precision grade and well calibrated.

ANNEXURE - III

Instrument Used – Company

Hot air oven	-	Yorco and Hirayama
Refrigerator	-	LG
Dona balance	-	ASIND (0.1 mg to 200g)
pH meter	-	Cyber lab
Weigh balance	-	Cyber lab
Millipore water	-	Direct Q
Centrifuge	-	Eppendrof
Dry Bath	-	Genei
Electrophoresis	-	Genei
DC output	-	Genei (50 V- 100V)
Spectrophotometer	-	Cyber lab
Bio RAD	-	UV doc